KLAUS TASCHWER, JOHANNES FEICHTINGER,
STEFAN SIENELL, HEIDEMARIE UHL (EDS.)

EXPERIMENTAL BIOLOGY IN THE VIENNA PRATER

On the History of the Institute for Experimental Biology 1902 to 1945

Klaus Taschwer, Johannes Feichtinger,
Stefan Sienell, Heidemarie Uhl (Eds.)

EXPERIMENTAL BIOLOGY IN THE VIENNA PRATER

On the History of the
Institute for Experimental Biology
1902 to 1945

Translation from German: Joanna White

AUSTRIAN
ACADEMY
OF SCIENCES
PRESS

Editors (for the ÖAW):
Klaus Taschwer, Johannes Feichtinger, Stefan Sienell and Heidemarie Uhl
Text, unless stated otherwise: Klaus Taschwer
Design: Armin Karner, Image processing: Otto Beigelbeck
Translation: Joanna White

Illustrations:
Cover: Front view of the BVA building in the Prater, photographed by Erich Smeikal,
Picture Archives of the Austrian National Library, 430.152-B
Back cover: Illustration from Hans Przibram (1930): Zootechniken. Experimental-Zoologie, vol. 7.
Vienna/Leipzig: Franz Deuticke, Plate 1.

Printed on chlorine-free bleached cellulose, acid free, non-ageing paper.

CONTENTS

Preface 6

Foreword 8

Experimental Biology in the Prater 10

The Changing Fortunes of the Address 'Prater No. 1' 12

Experimental Biology in Context Around 1900 15

An Idealistic Triumvirate in the Service of Research 18

Hans Przibram the Artist 21

A Unique Infrastructure 24

A Generous Donation to the Academy 27

The 'Toadkisser' and the Experimental Institute 30

The Scandal of the Nuptial Pads 33

Failed Careers at the University of Vienna 36

A Medical Star of the Interwar Years 39

The Co-Founder of Sex Hormone Research 42

From the Photo Album of the Institute for Experimental Biology 46

The Institute for Experimental Biology after the 'Anschluss' 50

The Fate of the BVA Staff 53

Destroyed in 1945 and then Suppressed 56

Re-Entering Collective Memory 59

A Short Chronology of the Institute for Experimental Biology 62

Selected Further Reading 66

Editors / Credits 68

PREFACE

With this publication and the exhibition it is based on, *Experimental Biology in the Prater: On the History of the Institute for Experimental Biology 1902–1945*, which opened on 12 June 2015 in the Aula of the Academy of Sciences (ÖAW) and was on display until 10 July of the same year, the Austrian Academy of Sciences honours the Institute for Experimental Biology in the Vienna Prater, one of the world's first research institutes for experimental biology.

In 1914 the imperial Academy of Sciences was presented the Institute for Experimental Biology (*Biologische Versuchsanstalt*, or BVA) by its founders Hans Przibram, Leopold von Portheim and Wilhelm Figdor as a gift. The donation not only included the building and its facilities but also an endowment fund worth 300,000 kronen, which had been set up by the brothers Hans and Karl Przibram with Leopold von Portheim and the income in interest from which was to secure the continuation of research activities. Following Karl Kupelwieser's endowment of the Institute for Radium Research to the imperial Academy of Sciences, the acquisition of the BVA was another decisive step in building the Academy's profile as a place for research in the natural sciences.

The 'Anschluss' in 1938 and the National Socialist seizure of power in Austria represented a radical break, in particular for the two institutes of the Academy of Sciences, the Institute for Radium Research and the Institute for Experimental Biology, at which numerous – male and female – scientists were carrying out research. Many of them had Jewish backgrounds and were removed from the Academy on 'racial' grounds. They were persecuted and driven out of the country. We know of eight people working at the Academy of Sciences who died or were murdered in National Socialist concentration camps: the Romanist Elise Richter and seven members and staff of the BVA – Leonore Brecher, Henriette Burchardt, Martha Geiringer, Helene Jacobi, Heinrich Kun, Elisabeth and Hans Przibram.

The treatment of members and scientists of the BVA and other Academy institutions who were persecuted as Jews during the period of National Socialist rule is a shameful chapter in the history of the Austrian Academy of Sciences. The Academy took the occasion of the 150th anniversary of its founding to engage critically for the first time with its role under National Socialism. In 1997 it commissioned Herbert Matis to publish a study entitled *Zwischen Anpassung und Widerstand. Die Akademie der Wissenschaften 1938-1945*. In 2013 the comprehensive volume *The Academy of Sciences in Vienna 1938 to 1945* appeared.

From this year, a plaque at the entrance to the festive hall at the ÖAW commemorates the 'Victims of National Socialism among the Staff of the Academy of Sciences'. Their names and fates can be looked up online in the *Memorial Book for the Victims of National Socialism at the Austrian Academy of Sciences*.

The BVA occupies a special position within the ÖAW's examination of its past during the National Socialist era. This publication hopes to provide the impetus for further, in-depth investigations into one of those Austrian research institutes which, over several decades, shaped the international academic landscape.

Anton Zeilinger
President of the Austrian Academy of Sciences

Photo: Sepp Dreisinger

7

FOREWORD

On 12 June 2015, the exhibition *Experimental Biology in the Prater. On the History of the Institute for Experimental Biology 1902–1945* was opened in the Aula of the Austrian Academy of Sciences (ÖAW). For the first time, the history of one of the world's first research institutes for experimental biology was the subject of a short exhibition featuring hitherto unpublished images. The following publication serves not only to document that exhibition but also to present the tragic history of this unique research institute in greater depth.

The Institute for Experimental Biology (*Biologische Versuchsanstalt*, or BVA), was founded by the biologists Hans Przibram, Leopold von Portheim and Wilhelm Figdor in 1902 as a private research institute and it was opened on 1 January 1903. To serve as its premises, the founders purchased the former Vivarium building in the Prater and invested their own money into turning it into one of the most state-of-the-art biological research facilities of its day.

Gift to the Academy

To secure the long-term existence of their research institute, Przibram, Portheim and Figdor donated the Institute for Experimental Biology to the imperial Academy of Sciences in 1914. The acquisition of the BVA and the Institute for Radium Research, likewise established with private funds, meant that from then on, the Academy was able to build a reputation as a research establishment in the natural sciences as well in the humanities.

Research at the BVA was interdisciplinary, international and carried out using cutting-edge equipment and laboratory facilities; all this made the institute the model for countless other research facilities from the Soviet Union to the USA. Leading figures in research worked there, including Hans Przibram, the co-founder of experimental zoology, Eugen Steinach, the pioneer of hormone research, and Paul Kammerer, whose controversial experiments to prove the heredity of acquired characteristics caused an international sensation.

Locked Out After 36 Years

The 'Anschluss' and with it the seizure of power in Austria by the National Socialists in March 1938 hit the Institute for Experimental Biology especially hard. Hans Przibram and Leopold Portheim, its two surviving founders, were locked out of the BVA from one day to the next. Those working there who were classified as 'Jews' by the Nuremberg Race Laws were also forbidden from entering their workplace after 13 April 1938. This lockout meant that for the two founders who had run the BVA for 36 years, the department head Eugen Steinach, and 15 employees, their research had come to an end.

Some of the members of the BVA persecuted on 'racial' grounds were able to flee the country; others were killed in the concentration camps. It is certain that seven BVA researchers died in National Socialist camps: Leonore Brecher, Henriette Burchardt, Martha Geiringer, Helene Jacobi, Heinrich Kun and Elisabeth and Hans Przibram. Their fates and those of – as things stand – 58 other Academy scientists who were victims of the Nazi regime are now commemorated online in the *Memorial Book for the Victims of National Socialism at the Austrian Academy of Sciences* (www.oeaw.ac.at/gedenkbuch – only available in German).

Research Comes to an End

With the dismissal of the BVA's staff, experimental research in biology in Vienna was over. In the final days of the war, the BVA building was largely destroyed during fighting in the Prater. All the materials and documents still housed in the BVA building, including Hans Przibram's unique specialist library for experimental biology, which he had been forced to leave behind in 1938, were destroyed by fire. All that remained was the correspondence between the BVA and the Ministry of Education, which is now kept in the Austrian State Archives. The Academy's central administrative files dealing with the BVA can be viewed in the archives of the ÖAW. Five archive boxes (around one metre of shelving of documents and images) is all there is to be found on almost three decades of

the BVA as an institute of the Academy of Sciences.

On 12 June 2015, the editors of this volume initiated and organised a day of commemoration at the ÖAW, which is documented in the final chapter of this publication. Academy President Anton Zeilinger together with Vienna City Councillor for Cultural Affairs, Andreas Mailath-Pokorny, and with several descendants of the BVA's founders present, unveiled a memorial plaque in the Prater Hauptallee. The plaque is located on the site where the BVA stood until the building was torn down. After the opening of the exhibition the Israeli evolutionary biologist Eva Jablonka gave an address honouring the continued relevance of research at the BVA. Beforehand, a bust of Hans Przibram was unveiled in the Aula of the ÖAW which had been given to the Academy in 1947 by Przibram's daughter, Doris Baumann, and his brother Karl Przibram.

The decision of the Academy's Presiding Committee to erect the bust in the Aula had never been carried out. With his bust now installed in the gallery among the Academy's greatest patrons, Hans Przibram – as one of its biggest donors and most important scientists – has been given a fitting place in the Academy's history.

**Klaus Taschwer, Johannes Feichtinger,
Stefan Sienell, Heidemarie Uhl**

EXPERIMENTAL BIOLOGY IN THE PRATER

The Institute for Experimental Biology,
founded in 1902, was one of the most important
research institutes in Austrian scientific history.
Just one of several good reasons why it is
fitting that we should remember
the history of this institute.

For over three decades at the start of the 20th century, biological and medical history was written at the Institute for Experimental Biology (*Biologische Versuchsanstalt*, or BVA) in Vienna's Prater. Nevertheless, few traces of the Vivarium's existence remain in Vienna. On the site where, 100 years ago, the research facility of international standing that was the Institute for Experimental Biology was located, there now stands a modest, one-storey building owned by the City of Vienna: the Road Safety School, bordered by pine trees and encircled by a miniature railway.

Until 2015, just a single street sign standing some 100 metres away recalled the long-vanished, sumptuous, neo-Renaissance building. In February 1957, Vivariumstrasse in Vienna's 2nd district was named after the building which was built in 1873 for the World's Fair and which housed the BVA from 1902 until, in April 1945, it was turned into a burnt-out ruin during the last days of the Second World War.

Good Reasons to Commemorate

The complete eradication of this highly innovative and internationally renowned research institute for experimental biology after the 'Anschluss' of Austria to Nazi Germany is one of the greatest tragedies of Austrian scientific history. No institute lost more of its researchers in the Holocaust than the Institute for Experimental Biology.

The former building of the Institute
for Experimental Biology with the Prater Hauptallee
in the foreground – photographed looking towards
the Riesenrad and Praterstern. Today a memorial plaque
on this site commemorates the institute.

The tragic end of the BVA and many of its staff is certainly not the only reason to commemorate the Institute for Experimental Biology. For one thing, the BVA was the most important research institute to be privately founded and financed by scientists in the history of Austrian science. Furthermore, in the early years of the 20th century, the Institute for Experimental Biology was one of the most important facilities for experimental biology in the world.

The BVA as a Cutting Edge Institution

The young botanists Wilhelm Figdor and Leopold von Portheim and the zoologist Hans Przibram invested substantial sums of their own private capital to purchase and equip the Vivarium. And they dug deep into their own pockets again in order to donate the BVA to the imperial Academy of Sciences.

The scientific importance of the institute was to biology what the Vienna Institute for Radium Research was to physics. At the Institute for Experimental Biology, not only were new fields opened up using totally new methods.

Its judicious directors also devised an innovative infrastructure and a new type of research organisation. Lastly, many of the topics researched at the BVA – 'epigenetic inheritance' deserves a mention here – have enjoyed a revival in recent times.

THE CHANGING FORTUNES OF THE ADDRESS 'PRATER NO. 1'

The building in which the Institute for
Experimental Biology was set up in 1902 had a
turbulent history. It was built in 1873 as an aquarium
for the World's Fair in Vienna's Prater,
later becoming a vivarium – putting itself not
least at the service of entertainment.

One of the more extravagant investments at the Vienna World's Fair, which took place in 1873 in the Prater, was something reported to be the 'largest aquarium in Europe': an ostentatious aquarium palace, set somewhat apart from the main pavilions. The exact location was to the south of the funfair (or 'Wurstelprater'), directly on the main thoroughfare, the Prater Hauptallee, which had been laid out already in the 16th century. Brought on board to design the new building with the address 'Prater No.1' was German zoologist Alfred Brehm (1829–1884), still known today as the author of the eponymous multi-volume work *Brehms Tierleben* (Brehm's Life of Animals).

From Aquarium to Vivarium

The enterprise was financed by Viennese industrial magnates who expected to make a profit on it. The spacious building was equipped with 16 large water tanks made of glass, which were lit only from above and were to be viewed from the dimly-lit aisles. The show exhibited both freshwater and saltwater animals: the saltwater was brought to Vienna by train from Trieste. However, both the World's Fair and, ultimately, the aquarium turned out to be loss-making enterprises – not least due to the economic crisis that began in 1873 and a cholera epidemic.

In 1887 the aquarium, under constant threat of bankruptcy, was bought by Karl Adolf Bachofen von Echt, a businessman with aspirations to educate the public and the mayor of Nussdorf (today a district of Vienna). He employed the Viennese naturalist Friedrich Knauer as the new scientific director, who replaced the word AQUARIUM over the entrance with VIVARIUM in gold letters.

The former aquarium was transformed into a sort of indoor zoo in which the fish had been replaced by a number of different animals: three orang-utans and a chimpanzee as well as countless types of birds. The former water tanks were used to house – in what were certainly not species-appropri-

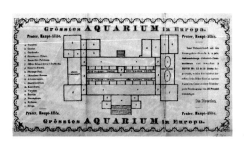

Fig. 1. Erster Dunkelsaal des Wiener Vivariums.

Detailed plan of the aquarium and the first dimmed room.

Plan of the World's Fair with the aquarium (blue circle) in the Prater Hauptallee.

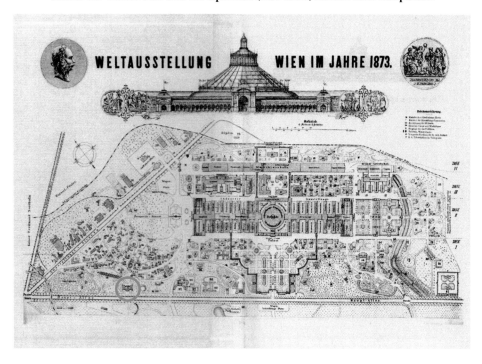

The aquarium's neo-Renaissance building in a drawing made immediately after the opening.

ate conditions – big cats such as lions, leopards, pumas and a panther.

Entertainment not Education

Plans by Knauer to set up a large zoo in the Prater under the auspices of the Vienna Zoological Society, founded in 1893, soon failed. Instead the focus returned to spectacle – the so-called ethnological expositions of the German zoo pioneer Carl Hagenbeck, for example. Meanwhile, the Vivarium was used by the Zoological Society from 1897 to exhibit the 'largest reptile collection in the world' – with 60 giant snakes, also supplied by Hagenbeck. Other acts to perform there included a female lion tamer.

In the winter of 1899/1900, a considerable number of the reptiles died and the now empty cages were filled with birds and small mammals. On 1 October 1900, the Vivarium was closed and the heating turned off; since most had come from tropical regions, this spelled the end for many animals.

In 1900, the zoologist Franz Werner feared that, in future, yet more entertainment would be needed to ensure the survival of the Vivarium: 'Yet more snake charmers, fire-eaters and parrot trainers will have to be deployed to attract visitors, and the whole thing will sink into a side show.' But come 1902, all that was about to change.

EXPERIMENTAL BIOLOGY IN CONTEXT AROUND 1900

As innovative as the concept
behind the Institute for Experimental Biology
in Vienna was, there were several international
and local developments that made its
foundation possible.

At the time of its foundation in 1902, the Institute for Experimental Biology was the first research institute to be dedicated not only to experimental biology, but to bringing botany, zoology and physiology, as well as neighbouring disciplines such as chemistry, together under one roof. No other facility had access to such a diversity of plant strains and animal breeds or had a similarly ambitious programme, encompassing as it did nearly all the big questions in biology.

In the Wake of Marine Biology

Before 1900, a whole series of developments took place in biology and in neighbouring disciplines that were necessary precursors to what was established in Vienna's Prater. One was the upsurge in experimental zoology in the final decades of the 19th century. This had been helped by the triumphal march of experimental methods in physiology, which marked the beginning of experiments on living organisms.

The breakthrough for experimental biology also followed in the wake of emerging marine biology. In the last third of the 19th century, research stations were set up in many of Europe's coastal towns. The most famous was the Naples Zoological Station, which opened in 1872 and which, thanks to its founder Anton Dohrn, soon became a stronghold of evolutionary and embryological studies. And it is surely no coincidence that is was after a long research sojourn in Naples that Hans Przibram took the establishment of his own institute in hand.

Another important prerequisite also turned out to be a non-university 'fad'. From the 1880s onwards, amateurs in Germany came together to form a number of aquarium and terrarium associations. The first association in Vienna was called 'Lotus' and one of its youngest members was an enthusiastic animal lover by the name of Paul Kammerer. On account of his success in breeding animals, he was

The main building of the University of Vienna,
where keeping live animals was prohibited in the period around 1900.

taken on by the BVA while still a student. But why was there no place for research into experimental biology at the University of Vienna? The immediate answer is a trivial one: a ban on the keeping of animals in the university building, which had been officially opened in 1884 on what was then the Franzensring, meant that pursuing experimental zoology there in around 1900 was out of the question. Moreover, the philosophical faculty, to which both the humanities and natural sciences belonged at that time, was suffering from massive underfunding.

Inferior University Lab Facilities

In 1902, i.e., the year that the BVA was founded, the natural scientists at the University of Vienna complained about the inferior standard of equipment in their laboratories in a 'Memorandum on the current situation in the Philosophical Faculty': 'Those who took part in the Meeting of Natural Scientists in Vienna in 1894 as a representative of a Viennese institute will not quickly forget the feeling of humiliation that came over them when showing foreign guests around the institutes.'

As a young student just turned 20, Hans Przibram also attended this scientific conference. One of the lectures was to determine his future path: 'At the Natural Scientists day in Vienna, Wilhelm Roux reported for the first time on developmental mechanics. At that moment, Przibram took the decision to continue Roux's line of research

The famous Zoological Station in Naples around 1900,
one of the models for the BVA in Vienna.

The library of the chemical laboratory
at the zoological station in Trieste, where Sigmund
Freud carried out research into eels in 1873.

AN IDEALISTIC TRIUMVIRATE IN THE SERVICE OF RESEARCH

The founding of the BVA was entirely
in private hands. It was initiated by the young
zoologist Hans Przibram and the two botanists
Wilhelm Figdor and Leopold von Portheim –
three wealthy scientists from Vienna's Jewish
bourgeoisie around 1900.

In contrast to many other countries, scientific patronage in Austria is limited to just a few wealthy donors. A good hundred years ago and the situation was a very different one. Members of Vienna's Jewish bourgeoisie in particular were keen to support research and, from the 1860s onwards, endowed scientific prizes (for example, the Lieben Prize) or financed research trips such as the Austro-Hungarian North Pole Expedition.

In most of these cases – including that of Karl Kupelwieser, the most noteworthy patron in the period around 1900 – these philanthropists were interested in science but were not themselves researchers. This did not apply to the founders of the Institute for Experimental Biology. The 28-year-old zoologist Hans Przibram and the two botanists Wilhelm Figdor and Leopold von Portheim, who were not much older, bought the Vivarium at the beginning of 1902 with their own money and built it up into one of the leading research centres for experimental biology in the world.

The Spirit of the Jewish Bourgeoisie

Spiritus rector of the enterprise was Hans Przibram, who was born in 1874 into a scientific dynasty from Prague that had many branches – similar to the Exners in Vienna or the Huxleys in England. Within the wealthy family, there reigned 'the spirit of the educated Jewish bourgeoisie of the liberal era', as physicist Karl Przibram (the brother of Hans) put it, a spirit 'open to all achievements in art and science'.

Among the many family ties and friendships enjoyed by the Przibrams was that of the Portheim family, also from Prague. Leopold Porges (Ritter von Portheim), born in 1869 in Prague, was Przibram's

Hans Przibram
in 1924
on his 50th birthday.

Wilhelm Figdor
in 1925.

Leopold Portheim
(seated, centre) in 1936
with the BVA staff.

19

**Hans Przibram
in younger years.**

cousin. Leopold von Portheim studied botany first at the University of Prague and then at the University of Vienna, where he was taught by the renowned plant physiologist Julius von Wiesner.

Portheim's wealthy background – his father Eduard Porges was vice president of the Prague Chamber of Commerce and had been awarded the noble title of Ritter von Portheim in 1879 – meant he saw no necessity for an academic career, seeing himself instead as an independent scholar. The establishment of the Institute for Experimental Biology, the financial side of which he was also involved in, created the ideal institutional environment for his botanical research.

The third member of this idealistic triumvirate of Viennese scientists was Wilhelm Figdor, who was born in Vienna in 1866 and, at 35, was the eldest of the group. Like several of the Jewish families whose palaces adorned Vienna's Ringstrasse, the Figdors numbered among those Viennese families who had risen to wealth during the second half of the 19th century. Already before the Institute for Experimental Biology was founded, Figdor had published a series of scientific studies from across the field of plant physiology.

Scientists as Generous Patrons

The three young scientists – and especially Hans Przibram – spent a total of around 300,000 kronen from 1901 onwards, first to buy the Vivarium and then to equip it with the most up-to-date facilities possible. In doing so, Przibram can justifiably go down in the history of Austrian science as the researcher to have put more of his own money into science than any other.

HANS PRZIBRAM
THE ARTIST

Science was not the only field
in which Hans Przibram was involved.
The zoologist possessed a remarkable talent
for drawing which meant that,
in the period around 1900, he became
something characteristic of Viennese
modernity – a traverser of
the border between science and art.

As a young student, Hans Przibram had a second passion besides zoological research: drawing. In 1895 and 1896, when a little under 20 years old, Hans Przibram produced accomplished comic strips in the style of Wilhelm Busch, then the most popular illustrator in the German-speaking world.

Przibram's early artistic works, which were only rediscovered in 2013, included an 'animal version' of *Shockheaded Peter* with a porcupine in the title role, a notebook on popular astronomy, city portraits from Europe and around the world, and an illustrated version of Darwin's *The Expression of the Emotions in Man and Animals*. These eleven exercise books full of comics were of such importance to him that he took them with him when he fled to Amsterdam in 1939. The books are still held there today in the city archives.

Pastiche of a children's classic:
first page of Hans Przibram's animal
version of *Shockheaded Peter*.

DIESES BUCH WURDE GEDRUCKT BEI
HERROSÉ & ZIEMSEN IN GRAEFEN-
HAINICHEN / DIE DRUCKANORDNUNG
WURDE GELEITET VON ADOLF LOOS
IN WIEN / DIE   CLICHÉS ZUM
BUCHSCHMUCK   VON STUD.
PHIL. HANS   PRZIBRAM
WURDEN VON   ANGERER &
GOESCHL IN   WIEN HER-
GESTELLT / VERLAG: GEORG HEIN-
RICH MEYER, BERLIN UND LEIPZIG /
LEITUNG DER REDACTION: STUD.
PHIL. ERICH VON HORNBOSTEL, WIEN.

Exhibition at the Vienna Secession

In 1900, on the recommendation of Adolf Loos, Hans Przibram provided over 200 vignettes and illustrations for a volume of poetry compiled by Viennese students (the *Musenalmanach der Hochschüler Wiens*). At the same time, the architect and publicist invited the young zoologist to participate in the winter exhibitions at the Secession in 1899/1900 and 1900/1901. Some of these works were published in the June 1901 edition of the magazine *Ver Sacrum*, the official publication of the Vienna Secession. His works were receiving favourable mentions in the British art magazine *Studio* as early as 1900 and the young Hans Przibram was also in demand as a designer of book plates.

Documenting the Research

Przibram's self-taught artistic abilities, his style greatly inspired by art nouveau, later proved advantageous to science – or rather, to the research at the Institute for Experimental Biology.

In later years, the zoologist furnished many of the BVA's publications, including several works by members of his staff, with illustrations. Even as late as 1930, Hans Przibram continued to emphasise the advantages of drawings for the illustration of biological findings over photography, whose artificial colours could not fully reproduce those found in nature.

MUSEN ALMANACH DER HOCH SCHUELER WIENS

In 1900, Hans Przibram provided illustrations for the poetry volume *Musenalmanach der Wiener Hochschüler* and an edition of *Ver Sacrum*.

The eleven exercise books filled with accomplished comic strips from the years 1895/1896.

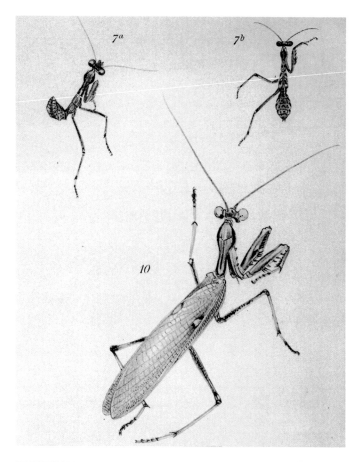

Przibram as a prolific scientific illustrator: sketches of a praying mantis, one of the model organisms at the Institute for Experimental Biology.

Colour-enhanced photographs of frogs from the Institute for Experimental Biology.

A UNIQUE
INFRASTRUCTURE

The innovative power of the BVA as a research
institute cannot be overestimated.
Not only was it equipped with the latest apparatus,
for example for researching the environment's
influence on animals and plants, but its
organisational structure was also exemplary:
research was interdisciplinary and international.

In general, scientific institutions are judged on how innovative the findings of their researchers are. In this respect, the Institute for Experimental Biology can also point to its countless discoveries over the 35 years of its existence, many of which are possibly waiting to be rediscovered today.

State-of-the-Art Technical Equipment

In the case of the BVA, much of the innovation was already there in how the institute had been set up: its technical infrastructure and the creation of artificial natural environments for biological research. But it was also there in the way it organised research. All this made the Institute for Experimental Biology a model for similar institutions, from Moscow to New York. Thanks to the financial clout of the founders, but also thanks to their impressive technical knowledge, extensive adaptations to the Vivarium building were carried out from 1902 and state-of-the-art equipment was brought in. In part these adaptations drew on existing architectural features, but they also saw the creation of individual workspaces, laboratories, stables, open air terraria and glass houses, six cement pools and a large frog basin in the grounds of the Vivarium.

The new temperature chambers in particular represented a pioneering achievement. They allowed experiments to be carried out at precisely controlled temperatures ranging from 5 to 40 degrees Celsius, as well as at a regulated humidity. This made it possible to study how animals adapted to different temperatures – including the pos-

Fig. 9.

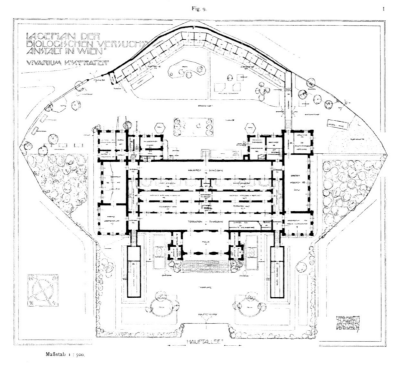

**Site plan of the Institute for Experimental Biology
showing the distribution of the various departments.**

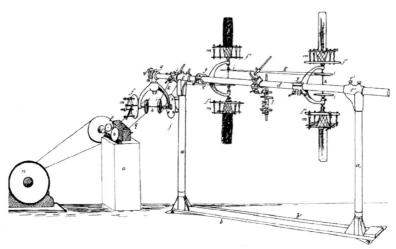

**State-of-the-art technology for the botanists: the so-called clinostat
was used to study the effect of gravity on plants.**

The chemical laboratory originally used by Wolfgang Pauli sen., the father of the Nobel Prize winner of the same name.

sible heredity of variations in colouration, for example, caused by heat or cold.

One special feature of the BVA's infrastructure was the immense variety of animals and plants kept and bred in the artificial environments. A list dating from 1908 shows 738 different species and subspecies from all classes of animal, including 101 types of mollusc (e.g. snails) and tunicates (e.g. sea squirts), 73 types of fish, 69 types of amphibians, 47 species of reptiles, 7 species of birds and 23 types of mammals. The BVA's zoologists were able to propagate around 40 per cent of these (297 in total), not least due to the expert knowledge of the young staff member Paul Kammerer.

It was not only the infrastructure but also the way research was organised that was highly innovative at the BVA. Scien-tists from different subjects (from zoology to physics and chemistry, from botany to physiology) worked together across different disciplines in order to answer the big questions of modern biology through experimentation. In addition, there was a lively coming and going of fellow scientists from Austria and abroad, who were assigned laboratory space for a period of time.

Breeding Ground of Modern Biology

All this took place under the careful direction of Hans Przibram in particular, who was mainly responsible for the fact that, especially in its early years, the Institute for Experimental Biology became a breeding ground for modern biology in Vienna.

A GENEROUS DONATION
TO THE ACADEMY

Following the model of the Institute for Radium Research
and the Kaiser Wilhelm Society, at the beginning
of 1914, after a lengthy period of negotiation,
Leopold von Portheim and Hans Przibram signed
over the Institute for Experimental Biology
to the imperial Academy of Sciences.

Until the end of the 19th century, scientific research in Austria was predominantly carried out at universities or state institutions such as the Natural History Museum or the Imperial Geological Institute (*Geologische Reichsanstalt*). The Institute for Experimental Biology was one of the first exceptions to that rule. In the early 20th century, two other innovative research institutes in the natural sciences were also set up: the Biological Station in Lunz am See in 1906 and the Institute for Radium Research in 1910, the first institute in the world dedicated to the study of radioactivity.

The lawyer and industrialist Karl Kupelwieser initiated and financed both institutes. From the beginning, the Radium Institute was under the direction of the imperial Academy of Sciences. This was also now the aim pursued around 1910 by the two founders who had remained active as directors of the BVA – Wilhelm Figdor withdrew from his leading role in the BVA during the handover –, not least from pragmatic and financial considerations. A further possible reason for the proposed incorporation into the Academy was the establishment of the Kaiser Wilhelm Society (*Kaiser-Wilhelm-Gesellschaft*) in Germany in 1911, the precursor to today's Max Planck Society, which gave a substantial boost to non-university research.

Research without Teaching

On 11 January 1911, the day on which the Kaiser Wilhelm Society was founded, Portheim and Przibram argued in a 'Promemoria' addressed to the Presiding Committee of the Academy that 'the thought of the benefits of our own research facilities, which, freed from the confines of university teaching, can devote themselves wholly to scientific research, gains ever more ground.' They cited the Radium Institute and the Kaiser Wilhelm Society as explicit examples of this.

The Academy decided to take advice on the matter and appointed a commission, which ultimately came to the conclusion that a research institute for experimental biology would meet a scientific and organisational need: 'Especially the modern

Handover document with which the BVA became part
of the imperial Academy of Sciences.

study of heredity and adaptation, the question of speciation, and much else besides require investigation through experiment. Extensive breeding trials at existing university institutes founder too easily on the facilities, while also quickly coming into conflict with the demands of teaching.'

The negotiations dragged on for several years as the Academy tried to rule out taking on any additional financial burden with the institute. In consequence, Leopold von Portheim and Hans Przibram each raised 100,000 kronen in securities, the interest from which would be used to facilitate the research programme. These

200,000 kronen represented a value of around one to two million Euros in today's money.

The Cherry on Top

Hans Przibram's brother Karl, a physicist at the Institute for Radium Research, increased this tidy sum by an additional 'reserve capital' of 100,000 kronen, to be used for any eventual renovations or new buildings at the institute. Now that this condition had been met, on 1 January 1914 the donation was finalised; it was, however, to bring little fortune to the BVA.

ad $\frac{1225}{1913}$ pr. 30./XII. 13.

Biologische Versuchsanstalt

in

Wien

II/2., Prater, „Vivarium".

TELEPHON 12857.

Wien, d. 29./XII 13.

An die

Kaiserliche Akademie der Wissenschaften in Wien.

Wir erlauben uns die Mitteilung zu machen, daß wir der k. k. Postsparkassa in Wien für das Konto Kaiserliche Akademie der Wissenschaften in Wien – Biologische Versuchsanstalt „Reservekapital" No 2.776.965 im Auftrage des Herrn Dr Karl Przibram

K. 20.000.– 4%. Mai – Rente m. Cp. 1/5. 14

fl. 20.000.– 4%. Böhmische Hypothekenbank Pfandbriefe mit Cp. 1/5. 14

fl. 20.000.– 4%. Mährische Hypothekenbank Pfandbriefe

überwiesen haben. mit Cp. 1/5. 14

Ferner wurden der k. k. Postsparkassa in Wien für das Konto Kaiserliche Akademie der Wissenschaften in Wien – Biologische Versuchsanstalt „Betriebskapital" No 2.776.966 im Auftrage des Herrn Prof. Dr Hans Przibram

K. 82.000.– 4%. Jänner – Juli – Rente m. Cp. 1/7. 14

K. 18.000.– 4%. Mai – November – Rente m. Cp. 1/5. 14

A total of 300,000 kronen of additional endowment capital
had to be raised by Hans and Karl Przibram and Leopold von Portheim
before the institute could be donated to the Academy.

29

THE 'TOAD KISSER' AND THE EXPERIMENTAL INSTITUTE

Among the Institute for Experimental Biology's many extraordinary members of staff, the zoologist Paul Kammerer was undoubtedly one of the most colourful figures. Traversing the border between science and art, his work in popularising science also made him a controversial figure.

O ver the years, a number of renowned life scientists carried out research at the Institute for Experimental Biology. These included not only its founders and its department heads, Wolfgang Pauli sen. and Eugen Steinach. Many great biologists also began their careers at the BVA – the later Nobel Prize winner and zoologist Karl von Frisch, for example, who wrote his doctoral thesis in 1910 under the supervision of BVA director Hans Przibram.

Probably the most colourful member of the BVA was Paul Kammerer, whose controversial experiments to prove the heredity of acquired characteristics became the subject of fierce debate and brought him great popularity. The public attention the young zoologist received was also bolstered by his own intense commitment to dissemenating scientific findings.

A Man of Many Talents

Kammerer was recruited during the founding phase of the BVA in 1902 as an unpaid assistant. The then 22-year-old had come to Hans Przibram's notice through his extraordinary ability to keep and breed animals. Kammerer had several other talents: he was a composer, had a flair for writing and was a *homme à femmes* – even managing to persuade Alma Mahler to come and work at the BVA for several months.

The young zoologist, who was awarded his doctorate in 1904 for a study on salamanders, began his own research programme at the BVA whose aim was to prove the heredity of acquired characteristics. His work on colouration change in fire salamanders as they adapted to the colours around them, and his claim that this acquired characteristic was then heredity, received much attention from both specialists and the general public.

Kammerer, whom Arthur Koestler memorialised in his 1971 book *The Case of the Midwife Toad* (in German: *Der Krötenküsser*, 1972), also successfully carried out other spectacular experiments. He induced the redevelopment of eyes in eyeless olms,

Paul Kammerer in November 1923 on his way to the USA.

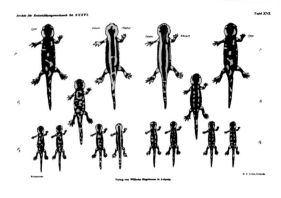

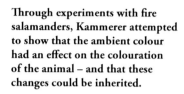

Through experiments with fire salamanders, Kammerer attempted to show that the ambient colour had an effect on the colouration of the animal – and that these changes could be inherited.

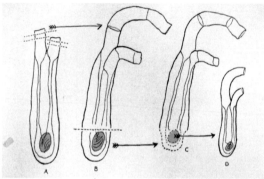

Kammerer also attempted to show the heritability of acquired characteristics in sea squirts.

for example, through the introduction of artificial light. With midwife toads, he was able to demonstrate inherited changes to breeding behaviour. However, the First World War brought Kammerer's experimental research at the BVA to an early end: nearly the entire valuable animal population, and with it Kammerer's breeding stock, perished.

Professional Populariser

When, in 1919, Kammerer's application to the University of Vienna for an unpaid, extraordinary professorship failed and when, after the First World War, he no longer thought it possible to survive on the wages of a BVA assistant, he gradually withdrew from experimental research and with it from the BVA, although he continued to collaborate with Eugen Steinach, for example. In the 1920s, Kammerer earned his living mainly through popular scientific lectures and books, which did little to lessen existing doubts about his experimental work.

After two long lecture tours to the USA in 1924 and 1925, Kammerer received an unexpected offer in 1926 to become a professor at the Soviet Academy of Sciences in Moscow. That he never went is due to one of the greatest unsolved criminal cases in scientific history.

THE SCANDAL
OF THE NUPTIAL PADS

Research at the BVA during the interwar years
was not only overshadowed by a constant shortage
of funds, but also by the affair surrounding Kammerer's
midwife toad – one of the biggest and, as yet,
still unsolved scientific scandals of the 20th century.

By 1926, the preserved specimen was already nearly 15 years old. But the midwife toad with nuptial pads, conserved in formaldehyde, had remained one of the most heavily debated objects in biology after the First World War. Between 1919 and 1926, over 30 articles that dealt with this controversial exhibit from Paul Kammerer's laboratory appeared in the renowned British scientific journal *Nature* alone.

What was it all about? Midwife toads are one of the rare species of frog that mates on land. After many years of breeding experiments involving raising the temperature in the terrarium, Kammerer was successful in forcing the toads into the water. There the male toads – as with most types of frogs – actually developed nuptial pads on their 'hands', which enabled them to hold on to the female more securely when copulating in water. According to Kammerer, this new characteristic was passed on by the midwife toad over several generations. For the zoologist, this was further proof of the heredity of (re)acquired characteristics.

An Influential Opponent

However, these claims soon found a bitter and influential opponent in the famous British biologist William Bateson, the founder of modern genetics, who left no stone unturned in casting doubt on the findings. Even a lecture tour to England in 1923, during which Kammerer showed the specimen to his British colleagues, could not overcome the geneticist's scepticism. As it was, the eccentric Viennese biologist had already long given up hope of an academic career by this time and had withdrawn from the Institute for Experimental Biology.

At the start of 1926, the young American zoologist Gladwyn Kingsley Noble from the American Museum of Natural History in New York travelled to Vienna with the express purpose of examining the disputed specimen, which BVA director Hans Przibram readily allowed him to do. During his inspection, Noble discovered that the nuptial pads had been manipulated with black ink. It seemed that Kammerer had been found guilty of scientific fraud.

A few weeks after Noble had published his spectacular findings in the scientific journal *Nature*, on 23 September 1926 Kammerer took his own life near Puchberg am Schneeberg. The reasons behind this act of desperation remain as unexplained today as the question of whether Kammerer really was the originator of the manipulations.

Loss of Reputation for the BVA

For the BVA, the international scandal, which was reported around the globe, was a heavy blow – not only because of Kammerer's suicide. The reputation of the BVA suffered massively, more so since Przibram, despite great efforts to solve the mystery himself, was not able to provide the Academy with any explanation for the falsifications.

It also helped little that Anatoly Lunacharsky, the Soviet People's Commissar for Education, sought to rehabilitate Kammerer in a play, which was then also made into a film in 1928 with the title *Salamandra*. New evidence suggests that Lunacharsky's supposition of a politically motivated, deliberate act of sabotage may not have been wholly incorrect.

The specimen of the midwife toad, whose 'hand' was manipulated with ink in order to give the impression of nuptial pads.

Kammerer's controversial defence: The Soviet-German silent film *Salamandra* from 1928.

The 'whistle-blower' Gladwyn Kingsley Noble
with microscope.

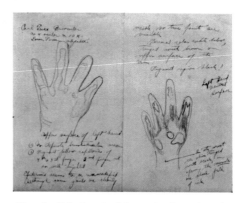

Sketch of the hand of the midwife toad made
by the American zoologist Noble,
who used this drawing to prove
the falsification in 1926.

Both the scandal and Paul Kammerer's suicide were the subject
of extensive reports in the American tabloid press.

FAILED CAREERS AT THE UNIVERSITY OF VIENNA

The Institute for Experimental Biology was considered, in particular by anti-Semites, as a 'Jewish' research institute. After the First World War, some of the BVA's scientists were denied university careers on 'racial' and political grounds.

One feature of the Institute for Experimental Biology – as was also the case for the Radium Institute – was the surprisingly strong presence of female researchers. Many of them earned a living as teachers but considered their main profession to be research, even when, after the First World War, this was only possible under very difficult conditions.

One of these scientists was the zoologist Leonore Brecher, born in the Bukovina, who worked at the BVA from 1915; in 1916 with Hans Przibram, and afterwards as his unpaid assistant. Brecher's main research focus was the question of whether and how colouration changes in animals (for example in the chrysalises of cabbage white butterfly) were conditioned by their environment and whether they were heritable.

'A very badly smelling people there'

Starting in 1923, she tried, with Hans Przibram's help, to be granted the teaching qualification (habilitation) for the University of Vienna – and failed in 1926. The main blame for this failure lay with the paleo-biologist Othenio Abel, who found in the decisive meeting that Brecher 'did not have sufficient authority to work with students.'

The arguments put forward by Abel, the coordinator of a secret, anti-Semitic clique of professors, were supported by the biologists Jan Versluys and Franz Werner – the majority. Later, while still in Vienna, the Germanophile Dutchman Versluys became a member of the Dutch *Nationaal-Socialistische Beweging* (National Socialist Movement). Werner, who became a member of the NSDAP (Nazi Party) in 1934, wrote a letter in late 1926 to an American colleague in which he said of the biologists at the BVA: 'It is a very badly smelling people there' (sic!).

As early as 1919, Paul Kammerer, at the time the most widely-known staff member of the BVA, failed in his application to the University of Vienna for an unpaid, extraordinary professorship. Here too Othenio Abel pulled the strings. In Kammerer's case, the official grounds for refusal were that he had written an 'all too popular' science book (*Das Gesetz der Serie* – The Law of Series). The real reasons were more likely Kammerer's left-wing and pacifist sentiments as well as his long carreer at the BVA.

In the same year as Brecher, another Jewish colleague from the BVA and one of

BERICHT

der Kommission betreffs Habilitation von Dr.Leonore B R E C H E R.

Die Kommission bestehend aus den Herren VERSLUYS,PINTNER,WERNER,
Hans PRIBRAM, ABEL, MOLISCH, DOPSCH hielt am 18.Juni unter dem Vor-
sitz des Dekans eine Sitzung ab, in welcher gemäss § 6 der Habilita-
tionsordnung über die persönliche Eignung beraten wurde.Nach eingehen-
der Diskussion wurde der Antrag von Professor Hans PRIBRAM auf persönliche Eignung nicht angenommen; die Abstimmung ergab 1 Ja, 3 Nein und
3 Enthaltungen.

Die Kommission formulierte den Grund der Ablehnung des Habilita-
tionsgesuches in folgender Weise.

Die Habilitation wurde abgelehnt weil die Kommission zu der Über-
zeugung gelangte,dass der Habilitationswerber nicht geeignet sei,den
Studenten gegenüber die für einen Dozenten erforderliche Autorität
aufrecht zu erhalten.

Die Formulierung wurde mit 6 Ja gegen 1 Nein angenommen.

Die Kommission beantragt demnach Ablehnung des Habilitations-
Gesuches von Dr. Leonore B R E C H E R.

Report on the application for a teaching post for Leonore Brecher, who was
refused primarily on account of her Jewish background.

The BVA team in 1923. Seated from the left: Hans Przibram, his assistants Paul Weiss and Leonore Brecher, both of whom failed to reach the 'Habilitationsnorm' – the standard for university teaching – in 1926, and Leopold Portheim.
Standing from the right: Auguste Jellinek and Theodor Koppanyi – like Weiss, both also emigrated to the USA.

Przibram's assistants, Paul Weiss, also failed to attain the teaching qualification – the habilitation – at the University of Vienna, despite his supervisor being able to present reviews from over a dozen national and international experts, who wrote in euphoric terms about the work of the 28-year-old.

Paul Weiss Emigrates

The undistinguished biologist Jan Versluys, who also had close private ties to Othenio Abel and who had Abel to thank for his promotion to a professorship, simply dismissed Weiss's theory as false – and with this, the case was closed as far as the professors of the philosophical faculty were concerned.

In 1927, Weiss left Vienna with a grant, emigrated to the USA and became one of the most influential post-war neurobiologists of his generation. In 1979 he was awarded (with Richard Feynman, among others) the National Medal of Science, the most important scientific award in the USA. One of his doctoral students was Roger Sperry, a Nobel Prize winner

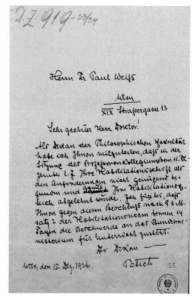

Letter from Carl Patsch, the dean of the university, to Paul Weiss about the 'official' grounds for his failed habilitation.

in 1981, who gained his doctorate writing about the very theory that had been Weiss's downfall in Vienna.

A MEDICAL STAR OF THE INTERWAR YEARS

He was nominated for the Nobel Prize eleven times
and was one of science's world stars in the 1920s and 1930s:
The physiologist Eugen Steinach co-founded
modern hormone research at the Institute
for Experimental Biology.

Today the name Eugen Steinach has largely been forgotten in Austria. Nevertheless, since 1955 the Steinachgasse in Vienna's 22nd district has recalled the once world-famous scientist, who conducted his most important research at the Institute for Experimental Biology and who, just under one hundred years ago, was one of the stars of international science.

From 1921 onwards he was nominated eleven times for the Nobel Prize for medicine or physiology and was often mentioned in the interwar years in the same breath as Sigmund Freud or Albert Einstein – as in this ironic poem written 1931:

"Drei Männer bilden das Staunen der Welt:
Der erste stürmte das Himmelszelt,
Der zweite der Seele Tiefen durchforscht,
Der dritte den alternden Leib entmorscht.

Und alle sind schon bei Lebenszeit
Todsicher ihrer Unsterblichkeit.
Was aber brüllt der alte Chor?
Die Juden drängen sich überall vor!"

'Three men have held the world in awe:
The first has stormed the firmament,
The second searched the depths of souls,
The third made aging bodies frail no more.

And all while still alive can be
Dead sure of immortality.
But hark, what roars that choir of yore?
The Jews are always pushing fore!'

Three Jewish stars of research in a contemporary caricature, 1931.

Eugen Steinach among his colleagues in front of the entrance to the BVA.

In these verses, whose ending is a dig at the anti-Semitism prevalent in Vienna, Steinach is celebrated as the scientist who made the human 'frail no more'. From his experiments on aging male rats, the physiologist and hormone researcher concluded that severing the ductus deferens – a vasectomy – would stimulate the body's own production of testosterone.

This procedure, which ultimately turned out to be ineffectual as afar as anti-aging was concerned, was known as the Steinach Operation and made the researcher famous in the 1920s. Personalities such as the Irish poet William Butler Yeats and Sigmund Freud underwent the procedure.

'To be Steinached'

Steinach achieved a still greater level of public fame with his *Steinach Film*. Produced with Austrian assistance, the documentary was premiered in 1923 at Berlin's UFA-Filmpalast am Zoo and presented Steinach's research in endocrinology to a wide audience.

'To be steinached' became a well-known saying. In Alfred Döblin's novel *Berlin Alexanderplatz* from 1929, for example, we find: '[…] Franz, you bash your forehead, what the heck's he done with himself, has he had himself steinached overnight? Aye, and then he starts talking and he can dance.'

A Foxtrot in Steinach's Honour

Karl Kraus mentioned Eugen Steinach several times in the *Fackel*; the physiologist was among Arthur Schnitzler's circle of acquaintances and was mentioned several times in his diaries. As early as 1920, the composer and music director Willy Kaufmann wrote the *Steinach Craze* foxtrot, and in the satirical magazine *Simplicissimus*, 23 articles refer to him and the 'rejuvenation craze'.

It was the *Steinach Film* in particular that captured the interest of the German pharmaceutical company Schering-Kahlbaum. In 1923, Schering's main laboratory began to research hormones and collaborated with Eugen Steinach, his assistant Walter Hohlweg and, from 1928, with the chemist Adolf Butenandt, who was later to win the Nobel Prize for Chemistry.

A great scholar strikes a pose: Eugen Steinach.

THE CO-FOUNDER OF SEX HORMONE RESEARCH

Eugen Steinach's scientific discoveries were overshadowed by numerous controversies and were largely forgotten after the Second World War. In recent years, there has been a rediscovery and revaluation of his pioneering work.

I t doesn't happen often in scientific history that an article written in German and published almost 80 years ago is published again in English translation. But that is exactly what happened in December 2013 with an essay published by Eugen Steinach and his colleagues Heinrich Kun and Oskar Peczenik in the *Wiener klinische Wochenschrift* in 1936 under the title *Beiträge zur Analyse der Sexualhormonwirkungen* (Contributions to the Analysis of the Effects of Sex Hormones), which has now been reprinted in the scientific journal *Endocrinology*.

Pioneer of Neuroendocrinology

The 1936 publication contains the first description of the role of oestrogen (i.e. the female sex hormone) in male rats, in particular in those which had been castrated. In 1972, oestrogen's special effects were discovered for a second time – with no reference to Steinach's publication of 36 years earlier. This long-forgotten essay by Steinach, so

argue chemists and historians of science in a commentary to the new translation, make the BVA physiologist at the very least a pioneer of neuroendocrinology.

Steinach's work in the 1910s and 1920s – for example, sex change in female rats through the implantation of testes – was seen as an important contribution to nascent sex science, which had just began to establish itself. From today's perspective, Steinach's suggestion that male homosexuals be 'treated' through testicle transplants is disconcerting. Yet Magnus Hirschfeld, the most prominent sexologist of the interwar years and co-founder of the homosexual movement, supported Steinach's claims at the time.

While Steinach's ideas of rejuvenation through vasectomy had already proved illusory by the late 1920s, his role as pioneer and trailblazer of hormone research is uncontested. Steinach's research not only led to Progynon, the first artificially produced hormone preparation, but to an extent it

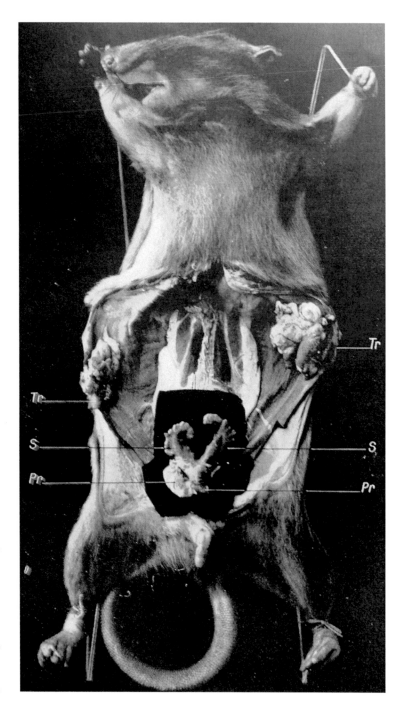

Rats in the service of hormone research: at the age of 1 month, this animal had testes (Tr) implanted, which remained in its body producing hormones for 14 months.

Co-developed by Eugen Steinach: Progynon, the first functioning hormone preparation. It was still in production until a few years ago and was used to combat the symptoms of menopause and in sex reassignment therapy.

was also a prerequisite for the hormone contraceptive 'pill', which has two Austrian 'fathers' in the physiologist Ludwig Haberlandt and the chemist Carl Djerassi.

Hormone Therapy for Livestock

Steinach's controversial yet celebrated hormone research was not limited to the effects of human sex hormones. To remedy infertility in cows, Steinach injected these animals with female sex hormones. His application of this method brought him financial success in particular in Switzerland and made him a pioneer of hormone treatment in veterinary medicine.

It was most likely his Swiss contacts that enabled him and his wife to find exile there after the 'Anschluss', while his villa in Böcklinstrasse was 'Aryanised' by the National Socialists shortly after the 'Anschluss'. Steinach, who was officially a researcher at the BVA until 1932, died in Switzerland in 1944 at the age of 83. It is possible that, even now, the pioneering significance of his work has not been fully recognised – in connection with the question of transexuality, for example, research into which was founded by Steinach's pupil Harry Benjamin in the USA only in the 1950s.

Above: One of Eugen Steinach's
colleagues at work.
Below: Steinach's comprehensive collection
of artistically preserved rat specimens.

FROM THE PHOTO ALBUM OF THE INSTITUTE FOR EXPERIMENTAL BIOLOGY

On the occasion of Eugen Steinach's 70th birthday,
the long-serving director of the physiological department
was presented with a photo album that documented
the BVA's facilities and his department.

Dedication in the photo album (composed and illustrated by Hans Przibram).

Biologische Versuchsanstalt (Gesamtansicht)

Physiologischer Trakt (Erdgeschoss)

External views of the Institute for Experimental Biology in winter 1930/1931: the main building and the physiological department, run by Eugen Steinach, as well as the outbuildings where Steinach's experimental animals – mostly rats and mice – were kept.

Tierstallungen

Outbuildings for animals
in the grounds and animal
cages inside the BVA.

Histologisches Laboratorium

Interior views of the BVA, in particular Steinach's department: it was possible to maintain the comparatively generously equipped laboratories thanks mainly to the cooperation with the German pharmaceutical company Schering-Kahlbaum.

Arbeitszimmer

Operationszimmer

THE INSTITUTE FOR EXPERIMENTAL BIOLOGY AFTER THE 'ANSCHLUSS'

In April 1938 disaster descended on this unique research institute. The directors were locked out and their property confiscated. In total, around two thirds of the researchers were barred from continuing their work at the BVA on 'racial' grounds.

The Institute for Experimental Biology is to be closed from 6pm this day for cleaning work that cannot be delayed and will remain closed until 25 April. At 8am on 26 April the Institute will be reopened to those members of staff who have applied for and been issued passes.' With these cynical words, the botanist Fritz Knoll, together with the designated Academy President Heinrich Srbik, issued an order which meant the de facto exclusion of the BVA's directors and Jewish staff members. Knoll had been appointed by the NSDAP as commissarial rector of the University of Vienna and charged with safeguarding 'the interests of the Nazi party' at the Academy of Sciences.

Expelled and Expropriated

This 'cleaning action', which also included changing the locks, was the final act that meant it was no longer possible for Hans Przibram and Leopold Portheim, who had been directors of the BVA for over 35 years and had donated it to the Academy, to enter the institute in which their life's work had taken place. It also meant the loss of valuable, privately owned equipment and a one-of-a-kind private research library. In addition, Fritz Knoll, who was responsible for the BVA at the Academy of Sciences, revoked Przibram's and Portheim's access to the BVA endowment fund, which they had set up.

Of the 29 people working at the BVA in March 1938, 15 came from Jewish families. They were all forced to leave the BVA. Another five went voluntarily in March 1938. With these departures, the BVA lost all its department heads and around two thirds of its staff.

In his report for the year 1938, the general secretary of the Academy portrayed these events in less dramatic terms: 'The Institute for Experimental Biology in the Prater is in the process of reorganisation in terms both of the design of the building and improvements to the interior fittings,

+ = Nicht-Arier

BIOLOGISCHE VERSUCHSANSTALT
DER
AKADEMIE DER WISSENSCHAFTEN
WIEN
II. PRATER, „VIVARIUM"

WIEN,

Liste der Arbeitenden 1938.

+	Brecher Leonore, Dr.	Zoologische Abteilung	Freiplatz
	Franke Ernst *ausgetreten*	*****	****
+	Geiringer Martha *ausgetreten 15.IV.38*	*****	*****
	Glaser Josef, Dr. *ausgetreten*	*****	****
+	Grünberg Friedrich, Dr. *ausgetreten 12.III.1938*	*****	*****
+	Häuslmayer Walter	******	****
+	Hausner Heinz, Ing.	Botanische Abteilung	****
	Hlavac Franz	Zoologische Abteilung	****
	Hrabik Ottokar *ausgetreten*	****	*****
	Jurisic Jovan	Pflanz.Physiol.Abteilung	*****
	Kisser Josef, Prof.Dr.	Botanische Abteilung	zahlend
+	Kurz Oskar, Dr. *wartet*	Zoologische Abteilung	Freiplatz
+	Kun Heinrich, Dr.	Physiologische Abteilung	*****
	Lenkl Katherine *ausgetreten*	Botanische Abteilung	zahlend
½+	Lindenberg Lise	****	Freiplatz
	Lohwag Kurt, Dr.	*****	****
	Mauser Franz, Dr.	Zoologische Abteilung	****
+	Peczenik Oskar, Dr. *ausgetreten*	Physiologische Abteilung	*****
+	Przibram Elisabeth	Zoologische Abteilung	****
+	Ried Oskar, Dr.	Botanische Abteilung	*****
+	Rix Karl *ausgetreten*	****	****
	Rzimann Gabriele, Dr.	*****	*****
	Schachl Josef	Zoologische Abteilung	*****
+	Schmidt Gerda	*****	*****
	Stift Alfred *ausgetreten*	*****	*****
+	Stock Alexander	****	******

Fortsetzung Liste der Arbeitenden 1937.

+	Traub Lorle *ausgetreten* Botanische Abteilung	Freiplatz
	Zeif Erhard	Zoologische Abteilung	*******
	Ziska Franz	Botanische Abteilung	*******

Abteilungsvorstände

+	Botanische Abteilung	Leopold Portheim
verstorben	Pflanz.Physiol.Abteilung	Prof.Dr.Wilhelm Figdor
+	Physiologische Abteilung	Prof.Dr.Eugen Steinach
+	Zoologische Abteilung	Prof.Dr.Hans Przibram
	Adjunkt	Ing.Franz Köck

The list of 'non-Aryan' researchers and departmental directors of the BVA (marked '+') in March 1938.

51

The seal of the Academy
of Sciences in Vienna between
1938 and 1945.

and the organisation of its scientific activities.'

The BVA ended in disaster. The research carried out by the newly appointed 'sub-authorised agent' Franz Köck, which mainly consisted of mixing sawdust with bran and testing this for use as livestock feed, proved to be nonsense. He was let go in autumn 1940, but not before he had brought about the destruction of the ponds and terraria, the closure of parts of the scientific collection and several showcases, and changes to the gardens, in short: 'severe injury to the value of the institute', as an internal report put it. Among other things, Köck's Nazi contacts had also enabled him to utilise Jewish forced labourers for BVA purposes.

In June 1943, the Academy of Sciences in Vienna concluded an agreement with the Kaiser Wilhelm Society which would allow the Kaiser Wilhelm Institute for Cultivated Plant Research to use the institute's facilities, greenhouses and gardens for its research. The war meant these plans never came to fruition.

And in the end, the building in the Prater returned to what it had been right at the beginning: a place for the display of fish in glass tanks.

THE FATE OF THE BVA STAFF

At least seven researchers who had worked
at the Institute for Experimental Biology until 1938
died or were murdered in the concentration camps.
No other research institute in Germany or Austria
lost more of its staff in the Holocaust.

The last that was heard from him was a postcard from Amsterdam dated 21 April 1943, on which he wrote a laconic message to his brother: 'Dear Karl! We have been summoned for travel to Theresienstadt …' A little over one year later, the life of Hans Przibram came to a tragic end: the highly-regarded biologist of international renown died in the Theresienstadt ghetto/concentration camp, probably of malnutrition and exhaustion. His wife Elisabeth committed suicide the following day. Just two of more than 33,000 people who went to their deaths in the Nazi's 'model ghetto'.

Failed Escape to the USA

Hans Przibram and his second wife had already fled to Holland together before the outbreak of war. On 3 March 1941, Przibram, an associate professor, wrote to Fritz Knoll, rector of the University of Vienna and the Academy's authorised agent for the BVA, asking him to send a letter in support of his planned emigration to the USA, where several scientists were prepared to support his departure from Europe.

The letter of support was written but, after consultation with the head of the Nazi Association for University Teachers, Arthur Marchet, the whole file was sent to Berlin, effectively rendering the Viennese letter invalid. In April 1943 the Nazis de-

**As the National Socialist rector
of the University of Vienna, as well as
the Academy's authorised agent for the BVA,
Fritz Knoll was partly responsible
for Hans Przibram's fate.**

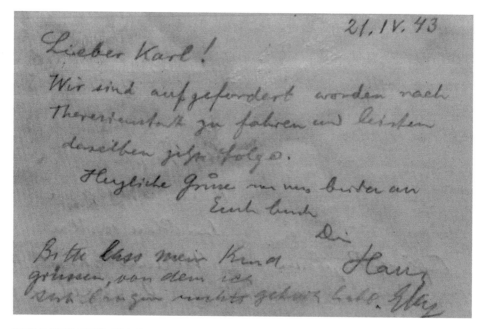

ported Hans Przibram and his wife to the Theresienstadt ghetto.

These two were not the only people working at BVA to die as a result of Nazi terror. In May 1942, Helene Jacobi was deported to the Belarusian extermination camp of Maly Trostinec near Minsk and murdered on the day of her arrival. Four months later, Leonore Brecher suffered the same horrific fate. All that is known of her colleague Martha Geiringer, one of Przibram's doctoral students, is that she

was deported from Belgium to Auschwitz in early 1943.

The physiologist Heinrich Kun, a close colleague of Eugen Steinach, died in an unknown camp in Yugoslavia. Henriette Burchardt, who also worked with Steinach over many years, was deported to Auschwitz in October 1944; the exact date of her death is unknown. The Institute for Experimental Biology was thus the research institute in Austria and Germany which, relative to its size, lost the most members to National Socialism.

1679 aus 1941/42.
M/T.

30.Dezember 1941.

An

Herrn Univ.-Professor a.D. Dr.Hans PRZIBRAM

in

Amsterdam-Holland,
Rijnstraat 162.

Mit Beziehung auf Ihr Schreiben vom 8.XII.1941
teile ich mit, dass ich Ihre Angelegenheit dem zuständigen Reichs-
ministerium für Wissenschaft, Erziehung und Volksbildung zur Entschei-
dung abgetreten habe.

Der Rektor der Universität
Wien:
In Vertretung

Hans Przibram's desperate attempts to emigrate to the USA failed in part due to a lack of support from those in charge at the University of Vienna.

Nationalsozialistische Deutsche Arbeiterpartei
Gauleitung Wien

N.S.D.-Dozentenbund
Der Dozentenbundsführer
an der Universität

An Se.Magnifizenz
den Herrn Rektor der Universität
Professor Dr. Fritz K n o l l

W i e n I/1
Dr.Karl Luegerring 1

Unter Zeichen: Doz/Ma/0319/3/41   Ihr Zeichen: GZ.14911/1679
aus 1939/40/41   Wien, den   19. März   19 41.
Betrifft: Prof.Dr.Hans Przibram,   I, Universität, Fernsprecher 21 500-72
Ausreise aus Holland.

M a g n i f i z e n z !

Der ehemalige a.o.Professor Dr.Hans P r z i b r a m
ist als Jude gegen den Nationalsozialismus eingestellt. Es
war dies schon zur Systemzeit daraus zu ersehen, daß er z.B.
gegen die Anregung, Professor A b e l bei seinem Abgang
nach Göttingen zu ehren, damals scharf Stellung nahm. Eine
Ausreise des Genannten in die USA würde ich trotzdem befür-
worten, da es wohl gleichgültig ist, ob er in Holland oder
in USA sitzt. Ich würde es sogar vorziehen, wenn Personen,
die gegnerisch eingestellt sind, aus Europa abwandern.

H e i l H i t l e r !

Der Dozentenführer
der Universität:

(Dr.A.Marchet)

1 Beilage.   Dozentenführer
                    d. d. Universität Wien

Rektorats-Kanzlei der Wr. Universität
Eingel. am 19. MRZ. 1941
G.Z. 1679   1 Beil.
Zu E. 14980, 1940/41

55

DESTROYED IN 1945
AND THEN
SUPPRESSED

The final days of the war in Vienna saw
the final destruction of the BVA building from which
the researchers had long been driven out.
After 1945, everything was done to erase
all memory of the BVA.

At the start of April 1945, SS units with heavy armoured vehicles were stationed in the BVA building. What happened next can be read about in the History of the Academy of Sciences, published in 1947 (on the occasion of its 100th anniversary): 'The building of the Prater Institute for Experimental Biology, II., Hauptallee 1, was destroyed by fire in the last days of the war.' This was followed by a short paragraph about the work of the BVA, but without a single word about the events post 1938 or even the briefest reference to the tragic fate of its staff.

'Biology for Sale'

The Academy – from May 1947 its official title was the Austrian Academy of Sciences (ÖAW) – wanted to be rid of the institute it had been given in 1914, which led to protests in the press. The newspaper *Neues Österreich* printed a final report under the title quoted above, commenting with resignation that: 'Certainly selfless

scientific curiosity has produced great achievements at other points in the city and the country, but at the heart of academia, in the traditional home of biology, material and even moral strength appear to have failed.'

One person who cut out and kept this and other newspaper articles on the sale of the Vivarium was the botanist Fritz Knoll, who must surely have felt this was directed at him. It is possible that the former Nazi rector, who was readmitted to the ÖAW in 1948, had a guilty conscience regarding the destruction of the institute entrusted to him in 1938 and the fate of the researchers at the BVA.

Missing Mentions

When, in 1951 and 1957, Knoll published two extensive volumes on famous Austrian natural scientists, physicians and engeneers, he had two opportunities to pay tribute to the Vivarium's scientists, especially as the BVA had been an Academy

The destroyed
and burnt-out Vivarium
building in 1945.

Weinhaus statt Forschungsanstalt
Eine traurige österreichische Geschichte

Die bürgerliche Presse berichtet, daß die Gründe im Wiener Prater, auf denen einst das Vivarium der Akademie der Wissenschaft stand, nunmehr für ein Vergnügungsetablissement verwendet werden sollen.

Das Vivarium war lange vor dem ersten Weltkrieg aus privaten Mitteln von den hervorragenden Forschern Hans Przibram, Wilhelm Figdor und Leopold Portheim begründet worden. Es war die erste Anstalt ihrer Art in der Welt und diente der quantitativen Erforschung der Beeinflussung der Lebensprozesse von Tier und Pflanze durch die Lebensbedingungen — niemals ein populäres Thema bei Reaktionären. Die Arbeiten aus dem Vivarium genossen Weltruf. So gehörte zu den Mitarbeitern der Anstalt der berühmte Hormonforscher Eugen Steinach, den die Ueberpflanzung von Geschlechtsdrüsen auch in breiten Kreisen bekannt machte. Dann der geniale Experimentator Paul Kammerer, der Vorkämpfer des Gedankens der Vererbung erworbener Eigenschaften, der durch die Intrigen und Machenschaften von Finsterlingen in den Tod getrieben wurde. Im Vivarium wirkte auch der hervorragende Kolloidforscher Wolfgang Pauli (Vater).

Während des ersten Weltkrieges nahm die Akademie der Wissenschaften das Vivarium in ihre Obhut. Die Nazi vertrieben die Gründer des Instituts und ließen es verkommen; die Bomben des Nazikrieges zerstörten auch die Baulichkeiten. Portheim und Steinach starben als Emigranten, Przibram wurde im KZ Theresienstadt in den Tod gehetzt. Das „jüdische Gedankengut" wurde vom Nazikommissär Köck vernichtet.

Eine andere Regierung als die unsere hätte es als selbstverständliche Ehrenpflicht angesehen, diese weltberühmte Forschungsstätte wiederzuerrichten. Statt dessen zwang sie die Akademie der Wissenschaften durch Entzug finanzieller Mittel, die Ruinen zu verkaufen, so daß das Gelände nun einem Zweck dienen soll, der den Figl, Schärf und Hurdes offenbar besser behagt als die biologische Wissenschaft.

Befremdend ist aber auch, daß in dem eben erschienenen Sammelband über Oesterreichs Naturforscher und Techniker keiner der Gründer und Mitarbeiter des Vivariums auch nur genannt ist. Ob die Akademie der Wissenschaften, deren Präsident doch ein aufrechter Demokrat ist, gut beraten war, die Herausgabe dieses Sammelbandes — der doch von nun an als offiziöses Nachschlagewerk verwendet werden wird — dem Parteigenossen Köcks, dem von der Wiener Universität entfernten Botaniker Knoll (Nazirektor im Jahre 1938) zu übertragen?

The Communist newspaper *Volksstimme* reported on
the BVA's final fate on 2 February 1951.

1952 saw the publication of
Sieg der Verfemten, a science-themed novel
based on facts about the BVA.

Arthur Koestler's book
The Case of the Midwife Toad (1971)
also recalled the BVA.

institute. But the principle of repression was stronger: Knoll wrote not a single word about any of them.

While the left-wing journalist Friedrich Lorenz may have written an astonishing fact-based novel about the BVA, which was published in 1952 under the title *Sieg der Verfemten* (Triumph of the Ostracised), and while the author and non-fiction writer Arthur Koestler produced a best-seller with his book *The Case of the Midwife Toad* (1971), which dealt not just with the alleged falsification scandal surrounding Paul Kammerer's midwife toad but also talked about Hans Przibram and his institute, it took almost four decades for the research institute, so long suppressed and forgotten, to be rediscovered.

RE-ENTERING COLLECTIVE MEMORY

The Second Republic came only very late
to dealing with the history of the BVA and the fate
of its staff. Since June 2015, a memorial plaque
commemorates the BVA at its location in the Prater.
A bust of Hans Przibram was also finally erected
after a delay of 68 years.

The memory of Hans Przibram, Leopold Portheim, Wilhelm Figdor and the BVA was almost completely erased after 1945. This is also evident from the history of the bust of Hans Przibram, which was unveiled in the Aula of the Austrian Academy of Sciences by Academy President Anton Zeilinger in a ceremony held on 12 June 2015. The academic sculptor Andre Roder had created the bust in 1933; in 1947 it was presented to the Academy of Sciences by Przibram's daughter, Doris Baumann, and his brother Karl Przibram. The Presiding Committee of the Academy decided to 'erect the bust in the Academy (in all likelihood in a niche in the Aula).'

No action was taken on this decision for a long time. Only 68 years later did the Austrian Academy of Sciences fulfil the wish of Doris Baumann and Karl Przibram for a marker of remembrance for the experimental biologist and scientific patron Hans Przibram. Next to the bust, a plaque bears the following text in German:

Hans Przibram (1874–1944) was one of the founders in 1902 and, until 1938, one of the directors of the Institute for Experimental Biology (BVA) in the Vienna Prater, one of the world's first research institutes for experimental biology. In 1914, the BVA was donated to the imperial Academy of Sciences, making the zoologist Przibram one of its most generous patrons. After the 'Anschluss' in 1938, he was persecuted on 'racial' grounds and relieved of his post by the Academy. Hans Przibram died in 1944 in the Theresienstadt ghetto/concentration camp. / Bust (1933) by sculptor Andre Roder (1900–1959) / Gift of Karl Przibram to the ÖAW in 1947 / Copy 2015, original in the Archives of the ÖAW.

On the same day, Anton Zeilinger, City Councillor for Cultural Affairs Andreas

Biologische
Versuchsanstalt
Vivarium

Hier befand sich die Biologische Versuchsanstalt (Vivarium), eine
der weltweit ersten Forschungseinrichtungen für experimentelle
Biologie. 1902 von Hans Przibram (1874–1944), Leopold von Portheim
(1869–1947) und Wilhelm Figdor (1866–1938) begründet und finan-
ziert, wurde sie 1914 der Akademie der Wissenschaften als Schen-
kung übergeben. Ihre Leiter Przibram und Portheim wurden nach
dem »Anschluss« 1938 »rassisch« verfolgt. Hans Przibram starb im
KZ Theresienstadt.

Here stood the Institute for Experimental Biology (Vivarium),
one of the world's first research facilities for experimental biology.
Established in 1902 and financed by Hans Przibram (1874–1944),
Leopold von Portheim (1869–1947) and Wilhelm Figdor (1866–1938),
it was donated to the Academy of Sciences in 1914. Its directors
Przibram and Portheim were persecuted on »racial« grounds after
the »Anschluss« with Nazi Germany in 1938. Hans Przibram died
in the Theresienstadt concentration camp.

| | Bust of Karl Przibram with his grandson Mathias Baumann (left) and Anton Zeilinger, President of the Austrian Academy of Sciences (right). | Memorial plaque for the BVA in the Prater Hauptallee, unveiled on 12 June 2015. |

Mailath-Pokorny and Mathias Baumann, grandson of Hans Przibram and there to represent the founders' families, unveiled a plaque to commemorate the Institute for Experimental Biology at its original location in the Prater Hauptallee. The bilingual plaque bears the following inscription:

Here stood the Institute for Experimental Biology (Vivarium), one of the world's first research facilities for experimental biology. Established in 1902 and financed by Hans Przibram (1874–1944), Leopold von Portheim (1869–1947) and Wilhelm Figdor (1866–1938), it was donated to the Academy of Sciences in 1914. Its directors Pzribram and Portheim were persecuted on »racial« *grounds after the* »Anschluss« *with Nazi Germany in 1938. Hans Przibram died in Theresienstadt Concentration Camp.*

Long Silence

Prior to this, the history of the BVA and the fate of its researchers had been ignored for decades in the Second Republic. The first comprehensive overview of the history of the Institute for Experimental Biology, written by the historian of science Wolfgang L. Reiter, wasn't published until 1999. The first symposium on the BVA wasn't held until 2002, organised by the Konrad Lorenz Institute for Evolution and Cognition Research (KLI).

**Descendants of the founders' families with Academy President Anton Zeilinger
and Vienna City Councillor for Cultural Affairs Andreas Mailath-Pokorny
after the unveiling of the BVA memorial plaque.**

2014 saw the publication of the edited volume *The Academy of Sciences in Vienna 1938 to 1945*, which portrayed 'the permanent destruction of the Institute for Experimental Biology and its academic staff' for the first time. That same year, on the occasion of the 100th anniversary of the BVA's donation to the imperial Academy of Sciences, the Austrian Academy of Sciences organised the symposium *One Hundred Years of the Institute for Experimental Biology*.

The conference was organised in co-operation with the KLI, which has been awarding a 'Hans Przibram Fellowship' since 1988. The conference programme was devised by Sabine Brauckmann. An edited volume in English is in planning. Further publications on the leading figures of the BVA have also been published in recent years or are forthcoming and document the growing interest in the scientific history of, but also the biological research that was carried out at this unique research institute. The biographies of the scientists who could no longer continue their research at the BVA after the 'Anschluss' in 1938 and were persecuted, expelled and/ or murdered, are documented online in the *Memorial Book for the Victims of National Socialism at the Austrian Academy of Sciences* (Gedenkbuch).

A SHORT CHRONOLOGY OF THE INSTITUTE FOR EXPERIMENTAL BIOLOGY

1873
Opening of the 'Aquarium' in the Prater Hauptallee based on plans by Alfred Brehm (of *Brehm's Life of Animals*) as part of the Vienna World's Fair.

1887
Under the direction of Friedrich Knauer, the aquarium building is officially renamed the 'Vivarium' and is now also used to exhibit amphibians and reptiles, later even large predators.

1901/02
The Vienna Zoological Society, which owns the Vivarium, is bankrupt. At the start of 1902, the zoologist Hans Przibram purchases the building together with the botanists Wilhelm Figdor and Leopold von Portheim. Costly adaptions to the building lasting several years follow.

1903
Official opening. Leopold von Portheim, Hans Przibram and their young colleague Paul Kammerer travel to Egypt and Sudan to collect living plants and animals for the BVA. Habilitation of Hans Przibram.

1907
Opening of the physical-chemical department under Wolfgang Pauli senior.

1909
Wilhelm Figdor is appointed Extraordinary Professor of Plant Physiology at the University of Vienna.

1910
Habilitation of Paul Kammerer and doctorate awarded to Karl von Frisch, who was supervised by Hans Przibram. Nearly all 20 staff positions are filled by

The BVA building over the course of its history: in the beginning was the aquarium (photograph taken around 1880).

post-doc researchers, some of whom now come from abroad.

1911
First proposals to donate the BVA to the imperial Academy of Sciences.

1912
Eugen Steinach takes over the directorship of the physiological department of the BVA and, over the following two decades, builds it into one of the world's leading centres for hormone research.

1913
Hans Przibram is appointed as an unsalaried Extraordinary Professor for Experimental Zoology.

1914
After lengthy negotiations, the BVA becomes part of the imperial Academy of Sciences. Wilhelm Figdor resigns from the board of the BVA. The physical-chemical department is disbanded following the departure of Wolfgang Pauli senior. Until 1916, the BVA is used as a hospital for wounded soldiers. In the following years, most of the animal stock dies.

1915
Limited resumption of scientific research.

1919
Kammerer's application for the title of Extraordinary Professor is denied by the University of Vienna. Proposals are put forward to move the BVA from the Prater to Schönbrunn. The following years are marked by severe financial difficulties.

1921
Hans Przibram is awarded the salary of an *Extraordinarius ad personam*.

**Until 1901 the building served as a vivarium
(photograph taken around 1900).**

1923

Kammerer is pensioned off. Walter Finkler's controversial study on transplanted insect heads is published.

1926

The American herpetologist Gladwyn Kingsley Nobel visits the BVA and discovers manipulations to Kammerer's legendary midwife toad, which leads to an international scientific scandal. Kammerer commits suicide shortly after publication. Leonore Brecher and Paul Weiss, both assistants of Przibram, fail to gain their habilitation.

1928

Reactivation of the temperature chambers.

1930

Przibram's monumental, seven-volume work *Experimental-Zoologie*, whose first volume appeared in 1907, is finally complete with the publication of the seventh volume.

1932

Steinach is awarded emeritus status. The public aquarium is reopened to bring in extra income. Proposals are put forward to open a laboratory to study the effect of radium on living organisms.

1938

Wilhelm Figdor dies before the 'Anschluss'. After the 'Anschluss', the founders Hans Przibram and Leopold Portheim are removed as signatories from the endowment fund and are literally locked out of the BVA – as are around half of the current staff. Leopold Portheim emigrates to London.

**In 1902 the Vivarium was transformed into the Institute
for Experimental Biology (photograph taken in 1931).**

1939

Hans and Elisabeth Przibram flee
to the Netherlands.

1942

The former BVA employees Leonore
Brecher and Helene Jacobi are murdered
in the Belarusian extermination camp
Maly Trostinec near Minsk.

1943

Agreement with the Kaiser Wilhelm
Society, which plans to use the BVA
building for the Kaiser Wilhelm Institute
for Cultivated Plant Research. Martha
Geiringer is deported from Belgium
to Auschwitz.

1944

Hans and Elisabeth Przibram die in the
Theresienstadt ghetto/concentration camp.
Eugen Steinach dies in exile in Swit-
zerland. The BVA researcher Henriette
Burchardt is murdered in Auschwitz.

1945

April: German combat units are quartered
in the BVA building during the final
battle for Vienna. Several fires destroy
most of the building and facilities.

1947

Leopold Portheim dies in exile
in London.

1948

The ÖAW sells the ruins of the Institute
for Experimental Biology.

SELECTED FURTHER READING

Baumann, Doris (1992): Dr. Doris Baumann. In: Dokumentationsarchiv des österreichischen Widerstands (ed.): Jüdische Schicksale. Berichte von Verfolgten, Vienna: ÖBV, p. 306–312.

Benjamin, Harry (1945): Eugen Steinach, 1861–1944: A Life of Research, The Scientific Monthly 61, p. 427–442.

Coen, Deborah R. (2006): Living Precisely in Fin-de-Siècle Wien, Journal of the History of Biology 39, p. 493–523.

Edwards, Charles Lincoln (1911): The Vienna Institution for Experimental Biology, The Popular Science Monthly 78 (37, 1), p. 584–601.

Feichtinger, Johannes, Herbert **Matis**, Stefan **Sienell** and Heidemarie **Uhl** (eds.) (2014, German 2013): The Academy of Sciences in Vienna 1938 to 1945, Vienna: Austrian Academy of Sciences Press.

Gaugusch, Georg (2011): Wer einmal war. Das jüdische Großbürgertum Wiens 1800–1938, A–K, Vienna: Amalthea.

Gedenkbuch für die Opfer des Nationalsozialismus an der Österreichischen Akademie der Wissenschaften, www.oeaw.ac.at/gedenkbuch/.

Gliboff, Sander (2006): The Case of Paul Kammerer: Evolution and Experimentation in the Early 20th Century, Journal of the History of Biology 39, p. 525–563.

Herrn, Rainer and Christine N. Brinckmann (2005): Von Ratten und Männern: Der Steinach-Film, montage/av, 14/2/2005, p. 78–100.

Hirschmüller, Albrecht (1991): Paul Kammerer und die Vererbung erworbener Eigenschaften, Medizinhistorisches Journal 26, p. 26–77.

Hofer, Veronika (2002). Rudolf Goldscheid, Paul Kammerer und die Biologen des Prater-Vivariums in der liberalen Volksbildung der Wiener Moderne. In: Mitchell G. Ash and Christian H. Stifter (eds.): Wissenschaft, Politik und Öffentlichkeit, Vienna: WUV, p. 149–184.

Kammerer, Paul (1926): Die Biologie in Wien, Urania 2 (10), p. 317–320.

Kassowitz, Max (1902): Die Krisis des Darwinismus. In: Wissenschaftliche Beilage zum 15. Jahresbericht (1902) der Philosophischen Gesellschaft an der Universität Wien, Leipzig: Barth, p. 5–18.

Koestler, Arthur (1971): The Case of the Midwife Toad. London: Hutchinson 1971.

Kofoid, Charles Atwood (1910): The Biological Stations of Europe. United States Bulletin of Education, no. 440, Washington: Government Printing Office.

Logan, Cheryl A. (2013): Hormones, Heredity, and Race: Spectacular Failure in Interwar Vienna, New Brunswick, NJ: Rutgers University Press.

Logan, Cheryl A. and Sabine **Brauckmann** (2015): Controlling and culturing diversity: Experimental zoology before World War II and Vienna's Biologische Versuchsanstalt, Journal of Experimental Zoology Part A: Ecological Genetics and Physiology 323 (4), p. 211–226.

Lorenz, Friedrich (1952): Sieg der Verfemten. Forscherschicksale im Schatten des Riesenrades, Vienna: Globus.

Müller, Gerd B. and Hans **Nemeschkal** (2015): Zoologie im Hauch der Moderne. Vom Typus zum offenen System. In: Karl Anton Fröschl, Gerd B. Müller, Thomas Olechowski, Brigitta Schmidt-Lauber (eds.), Reflexive Innensichten aus der Universität. Disziplinengeschichten zwischen Wissenschaft, Gesellschaft und Politik (= 650 Jahre Universität Wien – Aufbruch ins neue Jahrhundert 4), Göttingen: Vienna University Press, p. 355–369.

Przibram, Karl (1959): Hans Przibram. Neue österreichische Biographie, vol. 13, Vienna: Amalthea, p. 184–191.

Reiter, Wolfgang L. (1999): Zerstört und vergessen: Die Biologische Versuchsanstalt und ihre Wissenschaftler/innen, Österreichische Zeitschrift für Geschichtswissenschaften 10 (4), p. 104–133.

Södersten, Per, David **Crews**, Cheryl **Logan** and Rudolf Werner **Soukup** (2013): Eugen Steinach: The First Neuroendocrinologist, Endocrinology 155 (3), p. 688–695.

Taschwer, Klaus (2014): The two careers of Fritz Knoll. How a botanist furthered the Nazi Party's interests after 1938 – and sucessfully lived it down after 1945. In: Johannes Feichtinger, Herbert Matis, Stefan Sienell and Heidemarie Uhl (eds.): The Academy of Sciences in Vienna 1938 to 1945, Vienna: Austrian Academy of Sciences Press, p. 45–52.

Taschwer, Klaus (2014): Expelled, burnt, sold, forgotten, and suppressed. The permanent destruction of the Institute for Experimental Biology and its academic staff. In: Johannes Feichtinger, Herbert Matis, Stefan Sienell and Heidemarie Uhl (eds.): The Academy of Sciences in Vienna 1938 to 1945, Vienna: Austrian Academy of Sciences Press, p. 101–112.

Taschwer, Klaus (2015): Hochburg des Antisemitismus. Der Niedergang der Universität Wien im 20. Jahrhundert, Vienna: Czernin.

Walch, Sonja (2016): Triebe, Reize und Signale. Eugen Steinachs Physiologie der Sexualhormone. Vom biologischen Konzept zum Pharmapräparat, 1894–1938, Vienna: Böhlau.

Editors

Klaus Taschwer is a science journalist for the daily newspaper *Der Standard*.

Johannes Feichtinger is a Reader and member of research staff
at the Institute of Culture Studies and Theatre History at the ÖAW.

Stefan Sienell is the academic archivist at the ÖAW.

Heidemarie Uhl is a Reader and member of research staff
at the Institute of Culture Studies and Theatre History at the ÖAW.

With special thanks to Sabine Brauckmann,
Gerd B. Müller and Wolfgang L. Reiter.

Credits

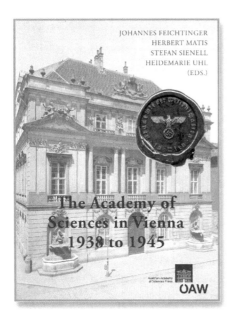

ISBN 978-3-7001-7537-7
Print Edition
ISBN 978-3-7001-7609-1
Online Edition
2013, 274 pp., 100 illustr.,
24x17cm, paperbound
€ 19,90

http://epub.oeaw.ac.at/7537-7

In 2013, the Academy is using the 75th anniversary of the "Anschluss" in March 1938 as an occasion to present an exhibition and to publish a catalog investigating the reactions of the Academy to the Nazi power takeover, the Academy's involvement in the Nazi domination apparatus, and the impact this had on the postwar period. The "Anschluss" of Austria by the National Socialist German Reich in March 1938 marked a profound turning point for the Academy of Sciences in Vienna. With the Nazis' seizure of power, Academy members and staff members were forced to leave for political and "racial" reasons. They were persecuted and expelled; they died in Nazi concentration camps. Under the Academy's new Nazi leadership, the learned society's autonomy was reduced and research projects in the support of Nazi ideology were carried out. The year 1945 was not a "zero hour." In addition to breaks, there were also continuities in the research institutes as well as the association of scholars. In dealing with Nazism, the Academy took an ambivalent stance: In the early postwar period, the membership of former Nazis was provisionally suspended. A few years later – pursuant to the Amnesty Law of 1948 – practically all former Nazi party members, even high-ranking officials, were re-admitted as members.

A-1011 Vienna, Dr. Ignaz Seipel-Platz 2
Tel. +43-1-515 81/DW 3401-3406, +43-1-512 9050
Fax +43-1-51581/3400
http://verlag.oeaw.ac.at, e-mail: verlag@oeaw.ac.at

AUSTRIAN
ACADEMY
OF SCIENCES
PRESS